TECHNOLOGY BRINGS WHAT NEGATIVE IMPACT TO

OUR SOCTIES

JOHN LOK

Contents

Foreword

Introduction

Nowadays, on the one hand, technology can bring beneficial positive impact to influence our life. e.g. automate manual driving vehicles, 3D product copying printer invention, advance medical equipment invention, space exploration fast speed shuttles or sky rockets invention etc. different advance technological equipment invention which can raise our standard of living quality and to pursue the aims of human further technological development. However, on the other hand, I also believe technology can bring negative impact to threaten our life in possible, if some scientists and/or these technological products manufacturers only consider themselves benefits to apply any these advanced technological invention to sell to any countries for weapon tools to help them to win competitive effort to achieve these countries can dominate or manage or control other pace countries pursuit.

In my this book, I shall give my opinions to explain why I believe technology can be applied to threat our life safety. I write this book aims to let scientists and these technological product manufacturer and the ambitious and dominant countries‘ leaders to consider that they have responsibilities to apply these technological products to bring any benefits to satisfy human's needs or enjoyment. Thus, any further scientific inventions ought only aim to be contributed to satisfy human's beneficial needs, they ought not to be manufactured to help any ambitious countries' leaders to raise their effort to apply these technological products to attack or dominate their enemy to encourage wars in our earth.

I shall concentrate on explaining these several technological invention aspects to let scientists to consider their further scientific inventions why will cause negative impact to threaten human's life safety if their pursuits are immoral to achieve their profit earning intention only from the ambitious countries' leaders. These important and influential scientific inventions include that, such as artificial intelligence, biochemistry medical research, nuclear energy invention, space exploration speed and alien communication pursuit technological invention.

Finally, in my this book, I shall give my opinions to attempt to explain why scientists ought consider these questions why their immoral intentions or behaviors will threaten human's life safety by these new technological inventions as below:

Can human become artificial intelligent machines' servants?

Can biochemistry healthy research be applied to become any diseases weapons to attack ourselves?

Can nuclear invention be applied to become different kind of nuclear bomb weapons to attack ourselves and to cause wars occurrence?

Is any space exploration expenditures be valueless or waste to carry on any unpredictable confident space exploration activities, such as Mar exploration or noon exploration etc.?

Will space technology become technological weapons to be applied to attack ourselves or to encourage space wars?

Ought scientists need to invent different technologies weapons to predict when space stones will fly to crash our earth to cause human death and to decide to apply what kind of technological weapons to attack the space stones, to replace to spend time and expenditure to invent any

telecommunication tools to communicate with alien?
This book is suitable to any readers who have interest to make judgement whether technology will bring negative impacts to influence human's safety in our future.I write this book concerns to research whether what technology negative influences will be caused to impact human quality of standard to be poor and brings negative consumer emotion. Although, it can rise productivity or performance to any entrepreneurs. But, technology can also bring negative impact to influence our life. Whether can it influence student learning abilities to be poorer? How can it bring negative impact to influence our life, e.g. poorer to influence student learning? I shall analyze it's negative impacts to our life on technological labor market, internet learning and computer entertainment and smart phone three kinds of technological aspects.
What are future business trends to influence any business development successful? What will be future potential development business? What are the nowadays business needs to adapt marketing change to satisfy client demand in the future? Any entrepreneurs need to consider whether what factors will influence their businesses to be succeed easily in global different industries competitive change environment.
In my this book, I shall choose some potential industries to let readers to feel which aspects who need to consider more to adopt their marketing future needs to satisfy future human living needs. I shall explain why some industry will be new trend to be developed and how to choose the best strategy to achieve to attract many clients' consideration for their new trend industry in order to win their competitors more easily. It is suitable to any readers who have interest to pursue their businesses can adapt to future

market change trends more closely.

It is suitable to any readers who have interest to research what factors will cause technology development to bring negative impact to cause human to get any disadvantages in our nowadays societies as well as how negative technology to any products can bring negavtive consumer emotion.

Prologue

Table of contents

CHAPTER I

Technology negative influence reasons

Online technology negative influence

Nowadays , internet is a popular tool to be provided to human to apply, e.g. online commerce brings businessmen to do online business trading, online searching information brings anyone can find information in short time, online studying can brings online learning chance and none classroom attendance to students. However, if we often do any online behavior, it will influence our mental and physical health to be poor, e.g. often spending time to use internet for social media contact. This interactive technologies will influence every young people's brain, behavior and attitude to be poor because they often spend time to use computer at home. Then, this digital technologies will lead them to lack nervous to study or learn any new knowledge, when who are students if they often apply computer to learn and they do not need to contact classmates and teachers in classrooms. Consequently, their school examination results will be possible influenced to be bad if they often apply computer to learn because they do not spend other time to any recreational activities or contacting people to make friends activities in their daily life.

It brings these two questions:

(1) Will internet often be used use to influence young people's mental and physical health to be poor?

(2) Has it bring direct negative impact relationship when young people often spend time to use interest to do learning and information research behavior to cause poor

mental and physical health?
Nowadays, human often uses internet which is one part of our habit. Our lives have become increasingly abuse in technology. Much of our communication and research is now online, much of our leisure and entertainment is provided by the internet and video games , and many of use internet find our mobile phones have become one essential part of our connectivity and everyday organizes to control our normal behaviors and to influence my normal life style poorly.
With these changes in lifestyle questions are it will arise negative influence about what technology may bring negative influence to us. Some of these questions bring potential detrimental effects, which had being unpredicted crisis in which the human brain is under threat from the modern world. Considerately, it influences the teenagers learning behaviors and attitudes to be poor. It seems that they have possible cause negative impact effect relationship between internet abuse habit behavior and poor mental and physical health as well as poor learning attitude and poor learning behaviors to young students.
The main factor of often doing internet playing behavior will have disadvantages to young people, because they will apply the internet tools to play video games to enjoy greater attention. This reflects a special case of environmental factor influence on whose mind and brain and health to be poor. Otherwise, if young people only spend some time to apply internet tools to do any reasonable need and meaning behavior, e.g. searching jobs from internet or searching any university written articles for study reference for learning intention or working seeking intention. Then, internet is a good tool to help them to develop their further career . Even, internet will train their brain and mind is more clear

and clever and health to get advantages during who do any searching behavior for studying or learning intention from internet channel.

In conclusion, internet communication technology will bring either positive or negative influence to any users, it is depended on the user how to spend whose time to do any researching data or studying behavior in their daily time spending arrangement . Such as often playing game behavior or watching movie behavior and listening music entertainment behavior , which will have negative influence to any internet users' mind and physical health to be poor. Otherwise, sometimes searching jobs or seeking teaching articles or newspapers to read for learning intention from internet tool, which will have positive influence to any internet users. Hence, internet technology must not bring negative influence to human, it can also bring positive influence to human. It is depended on how we spend time to apply this high technology communication tools to do the beneficial mind and learning training behavior from this technological communication tool.

- 1.2 How technology could contribute to bring poor standard of living to influence our societies

The effects of technology will have possible to bring global poor standard of living challenges. On the positive influence, especially science-based technology has offered a better world through the elimination of disease and material improvements to standards of living. But, on the negative influence, it will cause resource extraction, dangerous materials and pollution of air, water and oil have created conditions for unprecedented environmental to cause damage to the biosphere, when human applies any technologic tools to damage our earth natural environment in order to gain any profit for business aims.

Although technology brings businessmen to earn more profit, when who apply high technology to raise productivity and performance and efficiency to workers, e.g. artificial intelligence manufacturing robots, or they apply internet to sell their products (ecommerce), but technology also brings these disadvantages: Despite the ongoing technological revolution, the majority of the world population still lives in poverty with inadequate food, poor housing and less energy supply, illness increase , due to technological manufacturing can influence clean water and fresh air to be polluted to influence human's bodies to be un-health. Specially, the populations in Africa, Asia development countries, illness and death ratio both is risen by water and air pollution in these development countries nowadays.

Thus, it seems that it has relationship to bring negative influence to us between technology and air/water pollution and rising illnesses and deaths. Also, human needs to consider technology will support and enhance productivity and performance and efficiency , but it also influence human quality of standard to be poor challenge as the same time occurrence.

However, I suggest that businessmen ought reduce to invest much productivity by technological manufacturing improvement method, who ought concern environment pollution challenges how to avoid to apply technology to bring negative influence to all human's poor health challenge for long time. If human can apply technology, such as positive tool to solve problems or knowledge of how to create things, such as to brew beer , good taste soft drink or fruit or to make an atomic bomb, and culture

(or understanding of the world, our value-systems), e.g. agriculture , irrigation and clean water management and navigation technological skill improvement. It means knowledge, technology becomes understanding of how to make and use tools and instruments becomes encodes as technological knowledge and know-how.

Consequently, human's positive and responsible behavior will change technology tools to develop of modern scientific knowledge, based on observations, hypotheses and generalizations on the natural laws concerning the behavior of materials and the living environment.

How to avoid to technology brings negative influence on children

- Technology negative influence to children

In this world, it becomes impossible to escape the constant connection with others, aside from completely dis-connective from it, and into the unknown. Thus, parents need to know how their children using technological tools of behavior, which will influence impact on their children positively or negatively.

Nowadays, laptops and smartphones are now in the hands of children or young as ten age, and the eight to eighteen age young people that this group spends on average of ten hours and forty-five minutes or day exposed to media.

Whether their high amount of contact electronic media behavior is a good thing or not. So what is the right answer? Which side has the correct insight? When we may not have the immediate answer, one must look into both sides of the argument and determine what the correct path for today's children is. Thus, it brings thing effect, such as: one decision is about technology use will affect today's children as they develop.

Whether technology in classroom is truly a benefit for

students. The benefits include it can enrich basic skills. Students who have access to technology become more quickly in the material and , such as are able to absorb the information more quickly. Electronic material can be more stimulating and interactive for children, it is motivational since it provides ease to students in study conducted of advanced learning technology students have found to have more interested to attempt to do writing behavior.
Nowadays, children can use technology as a supplement with traditional education, but it is as not replacement. In fact, computers have been specifically useful, for they allow us to manipulate items, such as text to meet the needs of individual students. For example, text can be made larger so it can be seen easier and also read aloud for deaf students. Moreover, recently, specific devices have been engineers to cater to students with specific disabilities. Thus, it seems that the introduction of technology into modern culture has drastically shifted social norms to include technology into children's daily lives.

However, when technology had been applied essentially into children's daily live. Technology also had bad points. Today, it is not uncommon to bring children playing on their portable video game systems, when at a restaurant with their family or to see a child operating a computer better than some adults. If children were abuse to use computer to video game wherever they go to any places, such as restaurant, school, toilet, catching transportation tool to sit down to play video games by mobiles habitually. It will bring this social challenge: Can technology influence children choose not to pursue to spend much time to learn, instead of often spending time to play video games for entertainment aim by mobiles habitually.

Technology will part of word of the rest of our foreseeable lives. But if children often accustomed to apply technology tool to play any video games from mobiles and internet tool. Consequently, they will often devote nervous and time to spend to play any video games from mobiles conveniently any time. Just like there have to be rules of conduct in real life, there have not to be smart rule of conduct in digital life to children.

The pursue of this internet and video games entertainment technology will force or encourage children to the playing video games from internet skills to navigate it and keep up with it as they get old. Hence, to judge electronic media is beneficial or harmful to children's learning stage . It is depended on how the child chooses to apply computer and/or internet technology from electronic media tool. If the child often use internet and computer or mobile tool to go to anywhere to concentrate on playing video games. Then, I believe that it will bring harm to the child's future learning development. Otherwise, if the child often use internet and computer or mobile tool to learn or seek any education articles in classroom or library or at home, these electronic tools are as technology advances to learn media. Then , it will be beneficial to the child's future learning development.

CHAPTER II

Technology negative influence to low knowledge learner to feel difficult to adopt future new technological labor market

Nowadays, information technology development is rapid. It brings this question: Will it bring negative influence to low knowledge learner to feel difficult to adopt future new technological labor market, special in underdevelopment of culture countries' labor markets?

To answer this question, firstly, we need to know what the underdevelopment of culture countries' labor market means before to answer this question. Culture means adaptive behavior, has been an integral feature of the human species through its evolution, it is shared, learned, symbolic, and transmitted cross generationally. In another sense, culture refers to all non-biological aspects of human existence, including economics, politics and technology. Underdevelopment of culture countries' labor market means what labors are needed to the under knowledge or educational level countries' labor markets.

There are very strong beliefs that the adoption and usage of information technology has performed positive effects on the development of any country, but it is not present that it will can bring negative effects on the underdevelopment of the culture countries, e.g. Africa, island places' living people, these places are not reactive or are not reaching technology mature stage. So, it brings these questions:

- What if the rate of adoption exceeds society's or individual's ability to adapt, when the rapid introduction of information technology?
- What if economic benefits are distributed in ways that are socially destabilizing?
- What if income distribution is unfair, with higher skilled personal becoming better compensated, when many people are deskilled and effectively unemployed of jobs comparable to their current jobs and at salaries comparable to what they are earning today?
- Can their low knowledgeable workers feel difficult to learn any high technological skill to prepare their future job demand in these underdevelopment countries?

It seems rapid information technology to underdevelopment culture countries , which have chances to cause social challenges. Such as low skilled workers' unemployment , even office workers' salaries or technology manufacturing factory workers' wages will be reduced if who would not adapt the new technology development to follow the new technology influence to impact their work culture or method.

Thus, it also seems that culture can't exist without some form of society, i.e. culture us social. Therefore, cultural factors are observed in the society as providing to the production of its members who need to apply technological tools to manufacture products or serve their clients in their job responsibilities, e.g. factory workers, restaurant waiters etc. low skilled and learned workers. They need to learn how to apply new technology to work, e.g. computer skill or artificial intelligent skill.

So, it explains why rapid technology development will influence the low culture under development countries' low knowledge and low skillful workers to feel difficult to adapt how to learn to apply new technology production in themselves countries' technological job nature development change . Then , it will cause social challenges, such as unemployment, reducing wages, dismiss them, raising domestic labor market competition.

Moreover, some scientists concerned with the negative labor competition impact effect of information technology on the underdevelopment countries' low knowledgeable level of workers rather than the economic contribution of IT, because when many low knowledgeable level of workers feel difficult to learn technological skill to do their jobs, then they will be dismissed possible to bring social unemployment number to be increased and shortage of labor in these underdevelopment culture countries . In essence, they was asking if IT would erode this unique possession , even if it seems to contribute to their economic development.

Consequently, the intensive use of computer by the low knowledgeable skillful labors before the realization of whose thought –high technological production method itself may prevent the low knowledgeable workers' productive form being able to develop as a creatively thinking personally. This is a negative example of a mental process to them, which has been defined earlier.

In conclusion, the development of a new information society to under development culture countries would then raise a number of fundamental problems, one of which could be how to formulate and create optional cognitive preconditions for successful low knowledgeable labor' upgrade of high technological skill in short term. There are some pf the problems faced by developing nations who are still to development their technological production skill to low knowledgeable workers to let them feel difficult properly, let alone creating optimal preconditions for a successful mental process of low knowledge labor-computer interaction. At present, the adoption of IT in developing counties needed to be concern how to adapt whose countries' information technological labor users' production skill change.

The negative impact of smartphones/ mobiles and desktop/ laptop on human health and life

● 1 Avoidance to driving and speaking mobile at the same time

Nowadays, the smartphones being a very new invention of humanity, became an inherent part of human's life. The smartphone combines different features. It allows users to keep pictures, memories, personal information correspondence, health and financial data in one place. Smartphones also become an integral part of modern telecommunications facilities. In some regions of the world, they are the most reliable only

of available places. The phones allow people to maintain continuous communication without interruption of their movement and distances. However, recent scientific facts and research analysis of the smartphones' usage has disadvantages to influence human health and life.

The main key points indicate the effect of electromagnetic waves on human brains, effect of handheld device usage on human's upper extremities, back and neck. A significant neglect influence between the total time spend using mobile device each day and pain in the right shoulder and between times spent internet browsing and pain at the base of the right thumb. Moreover, mass cellphone calls enhance risk to human safety, e.g. when they are driving and listening and talking to touch mobiles at the same time. The drivers' driving and phoning calls behavior at the same time which will be very dangerous of their speaking and driving to cause traffic accident occurrence in possible. Thus, drivers can not neglect to avoid to do the mobile speaking and driving behavior at the same time when they are driving to reduce their traffic accident occurrence to cause their death or hurt in possible.

● 2 What are the negative effect of electromagnetic waves on human brains from smartphone influence

Scientists proved that the smartphone is a source of the eminence of electromagnetic waves. Numerous studies have been conducted in the past years to identify the effect of electromagnetic waves emitted from the cell phones on human health.

However, it has not proved smartphone can influence our health certainly. As soon as mobile phones more and more part of our lives, the world is continuing research to prove whether cell phones are harmful to human health.

Today, there is no official statement announced by laboratory or medical center to answer this question: The complexity of the analysis of the statistical data makes the task more difficult for researchers. The impact of harmful radiation emitted from cell phones is still being studies.

Nowadays, human are accepted to use mobile phone in any time, any where popularly. Although, mobile is a good small size and convenient carrying of communication tool for human to use when we need to make phone calls to anyone in anywhere and any time conveniently. But, I feel that it will harm human health when we often use this communication tool any time.

However, some doctors indicate cell phones can cause brain cancer risk easily. But they have not any evidences to prove it is truth nowadays. Hence, the statement that cell phones can cause cancer has been not confirmed. The studies failed to prove that cellphones make a major risk develop cancer among frequent users. The main issues when conducting studies are some people may not accurately report the usage as they don't exactly remember how often they use the cell phone excluding speaker phone , and it is still difficult to measure the impact of other factors that may accelerate the cancer development for excessive cell phone users.

Although, it is not proved that cellphone can use brain cancer to human when we often use. But some scientists or medical professionals have proved that the cell phone users often use cell phones , it is possible to cause human physical illnesses, such as upper extremities, back and neck caused unhealthy and pain.

A smartphone or handhelds device combines advanced computing capability, such as internet communication, information retrieval, video, e-commerce and other features, that make device highly popular among people. According to Pew research center investigating, it showed that the number of smartphone owners comprises 56% of American adults in 2013 year and their average daily use of the device is about 195 minutes. The number of cellphone users increase every year. Various studies show the connection between cellphones usage and physical illness of the users' health. Some studies report that users complain about a headache, hand tremor and finger discomfort and pain of physical illnesses numbers increasing.

In fact, most mobile hand-held device users complain of discomfort at least on one area of upper extremities, back or neck. Long –term usage of the device leads to additional tension on tenders , muscles and tissue etc. different kind of physical illnesses. Moreover, in research conducted by a group of Korean scientists from Inji University focused that an effect of cellphone on hand-held device users was a significant association between the total time spend using a mobile device each day and pain in the right shoulder, and between times spend internet browsing and pain at the base of the right thumb.

● 3 The laptop and desktop negative influence

On the laptop and desktop negative influence aspect, although telecommuting and telework communication technology is popular to be applied to our daily life. For example, they are modern alternative to office arrangement, employees work from home office, café, garden, carpark , even car.

According to scientists showed that nowadays, there are 20 to 30 million people who work from their home at least one day each week. Another 15 to 20 million work when they are on the road, 10 to 20 million runs some form of home business and 15 to 20 million work at home part of the time. IN most cases, people use desktop and laptop in their home office.

However, modified cellphones or smartphones are also substitutes to a home office. In fact, in principle of computers, it makes the workplace safer and convenient to compare mobile phones. There are different examples of adaption desktop or laptop computers to health needs of these users when bring their computers to go to anywhere to use in common, e.g. ergonomically designed keyboards design, pad bolster, mouse etc. design to adapt to their carrying to use their laptop or desktop needs. However, laptop or desktop computer products have not proved any serious harmful to influence human health to compare mobile phones at this moment.

Consequently , although technology can create different kind of jobs to let human to do, or assist human to communicate conveniently, e.g. mobile or artificial intelligent robot assist human to do any clerical job duties or learning more easily, .e.g. internet or laptop or desktop or owning mobile and laptop function computer products. There high technological products can bring benefits to satisfy human needs, .e.g. raising productivity efficiencies for workers, providing far distance overseas phone calls telecommunication, searching data or electronic business running from internet channel. But human can not neglect that these high technological products whether will bring negative influence to our mental or physical health when we often use them in possible. Thus, often using high technologies products to influence our health issue will be one important matter to be our future consideration.

CHAPTER III

What kinds of technologies innovation products will impact our future lives

Europe in the 21 St Century is a technological society, how today technological trends could impact upon society in ways to be fully considered by clients‘ needs. What technological advancement products which can carry trend with it the promise of saving time, or assisting business or manufacturer industry clients to do more in the same amount of time.

In our clients buying choice view point, who ought hope any technological innovation products which can offer them that the opportunity to do things more efficiently. I shall suppose that technological innovation will be the main factor which can attract future many clients' purchase choice from the owned technological innovation product seller. For example, mobility, resource security , electronic government technological innovation products will be popular trend in future technological innovation product market.

● Autonomous automatic vehicle

Can autonomous vehicles be popular in the future driving market? Will your child soon be driving you to work? The autonomous vehicles (artificial intelligent vehicles) will change the responsible driver concept. Why does autonomous vehicles will be future popular driving tools?

In fact, autonomous vehicles have these feature characteristics to differ to compare our common traditional driving tools. Their characteristics, such as real-time human control option, advantage of the large amount of high -quality mapping data of possesses to programing travel routes, exploring ways in which autonomous vehicle technology can be integrated with existing parking infrastructure to produce " driverless parking systems" accessible via existing personal electronic devices, e.g. smartphones is demonstrating the use of fully automated road transport systems in Europe and developing guidelines to design and implement such systems.

With some analysts predicting that by 2022 year , there will be around 1.8 billion automotive machine -to-machine connection its is clear that a large amount of data will be generated by vehicle in the future. Thus, this level of communication between automated vehicles should make to possible for such vehicles to navigate to destinations and interact with other vehicles and objects most effectively than a human brain. Moreover, they believe the chance of automatic vehicles' highway accidents occurrence will be less than traditional human driving vehicles.

Thus, the increased connectivity required to facilitate automation of vehicles would significantly improve the degree of monitoring of the performance of such vehicles. Individual owners would be able to better maintain and enhance their vehicles with improvements in fuel efficiency and lesser fuel spending and safety. This could also provide further benefits, such as terms of reducing traffic jams, reduced pedestrian exposure to pollution and lower risk of road-traffic and pedestrian incidents occurring, particularly in urban areas.

The rise of autonomous vehicles is also likely to combine with continuing electrification of vehicles as telecommunications software and hardware and further integrated into vehicles. Thus, the rental-orientated and purchase-orientated automatic vehicle business both models will have chance to be raised in future global driving market.

It causes the responsibility tends to lie with human drivers of vehicles will be decreased. A new set of IT skills in addition to a practical ability to drive and operate a more digital type of driving machine as well as it might impact upon existing vehicle users in terms of requiring re-training, particularly those less able to learn. Even, future public transport will have possible to be changed from non-human driving and change to automatic vehicle market will be individual and business both client markets in possible.

In conclusion, to success to sell non manual driving tools. The non manual driving sellers need to know how to solve these two artificial intelligent vehicles innovation questions: Could our future living habits change as a direct segment of changing transport behaviors? Will autonomous transport simply become and essential transportation tools for our homes and workplaces? Thus, if manufacturers want artificial intelligent vehicles sale number increases, which needs to influence future whose clients to accept this kind of non -human driving tools can be satisfy to change their traditional driving living habits for their new habit of non -manual driving method to substitute traditional manual driving tools.

- 3 D printer

Can 3 D printer be popular sale to manufacturing industry clients? What could be the effects to the physical environment and human health of such application 3D printer to copy to manufacture any productions? For example, medical equipment products, car keys, guns, furniture etc. different heavy or light weight manufacturing products.

The benefits to 3 D printer include: less production time, reducing purchase bulk or materials to produce any products, reducing to employ worker number to produce products, workers can learn to use 3D printer to copy to manufacture any products easily, to avoid air or water pollution to pollute working environment to influence worker health and safe production in factories, employers can pay less wages to employ less workers, workers can also raise more efficient during using 3 D printers to manufacture any products.

Thus, in the future 3 D printers can be popular to be used to copy to manufacture for these any products, e.g. jewelry or weapon industry products. In fact, 3 D printer is an additive manufacturing technology for making three-dimensional object, of almost one sharp using a digital model. Such as jewelry manufacturers apply it to copy to manufacture new kind of jewelry, hospitals can apply it to copy to manufacture any new medical equipment, weapon manufacturers can apply it to copy to manufacture any new gun weapons, aerospace or air plan manufacturers can apply it to manufacture new air plane engineering equipment or space exploration equipment or transportation tools. Thus, 3 D printer application will be popular to different aspects of manufacturing industry.

Future expected impacts and development for 3 D printer development. A macro economy level impact of 3 D printing will be considered to manufacturing industry business consumer-based economy and the societal behavioral acceptance in factories and offices manufacturing environment.

However, buying habits as individuals are able to print their own products, in comfort of their own home. Activity would be changed from traditional shopping methods to purchase 3 D printer to copy to manufacture own same products at home. Consumers can also choose how to design to print the product, rather than the manufacturing process itself is what consumers will be paying for and thus these is the potential for a design -lead choice behavior. Manufacturers don't need to buy many materials to manufacture products, they can use 3D printer , such as individual manufacturing machine parts, which could drastically improve their ability to design and manufacture more effective machine and components.

In conclusion, how to sell 3 d printer successfully. 3 D printer sellers need to know what advantages can give to 3 D individual consumption buyer and business buyer to let them to know to aim to let them to accept to change their buying behavior and manufacturing behavior for some products. There are some questions for consumers to attempt to answers:

What will the implications be the level of personal interactions between individuals in society of all of our products were to be manufacturing at home?

How would this change our typical buying habits and what would be the impact on our economy?

Would an increased use of 3D printing technology in the home or factory accelerate this process and what would be the implications for local high streets?

Would economies change being-focused will digital design skills having a greater benefits than traditional manufacturing methods?

If the ability to print everyday items at home becomes a reality , who is society would have the greatest access to

such technology?
If a particular demographic section (age, gender, race, income levels can be in factor to influence 3D printer consumer group, e.g. the 3 D printer buyer needs skills to manufacture any products, it seems only represented in a younger demographic. Could this mean that older members of society would not be able to benefit from 3 d printed projects?
In micro economy view point, although 3 D printer has benefits to individual and manufacturing consumers to reduce that their shopping or manufacturing expenditure, more design choice, raising worker individual skill and work performance. However, in macro economy view point, it also bring disadvantages to society. For example, if some members of society could not work move quickly, as a result than others, then what might be the impact upon their employability , e.g. causing unemployment of the 3D printer skillful learners who can not upgrade their working skill. Then, their employers will choose to dismiss these low skillful level 3 D printing learning skillful workers. Consequently, it will cause these member group of worker unemployment in the future society in possible. In conclusion, employers can not neglect how to train workers to learn how to apply
3 D printing skills to copy to manufacture any products.

- Massive open online course education

Will online education change traditional education? Basically, the students who choose to study from online channel, who must need have personal computers at home or school and often use internet from online platforms. In contrast to traditional methods of teaching with much small class size because every student can learn from online course at home. It means one teacher can choose to teach only one student from online channel. So, the teacher can stay at home or school as well as the student can stay at home , both of them can teach and learn from online teaching platform at the same time.

Whether the primary school, high school and university students who can accept to choose their learning habit to learn from this kind of online learning method more easily. In fact, online education will resultant impact on any teaching competitiveness. Due to , it is attempted to develop one kind of new technological education method to replace the traditional classroom by face -to-face teaching method between teacher and students contact.

However, it is not all course are suitable to adopt online teaching method and some courses re pointedly directed towards areas of interest that help education providers to also sell other online course products what other simply promote passive learning. For example, music, art, history, math, commerce courses which can be taught by teacher from online channel more easily. Because they do not need students to go to laboratory to do any experiments. Otherwise, engineering, food science, space science, medicine , doctor courses which need students go to laboratory to do experiments often. Thus, they are not suitable to be taught by teacher from online channel. Classroom teaching is more suitable to them.

Although, online teaching is low cost , due to that schools do not need many classrooms, even employ many teachers. So, they only buy computers and provide online education and less teachers are employed to teach whose students. So, it brings this question: Simply coursing cost barriers of success to education would not necessarily result in automatic take-up by young student consumers. May also need to think about best to education market, particularly to disadvantaged groups , such as older generations with lower computer and internet skills.

Who would be the winners and losers of an education market based upon such stronger principles of knowledge sharing and how can the institutions employing the use of such online or classroom or distance learning education methods be appropriately supported to maintain the high quality of further education? It seems to persuade students to choose online learning, the only method is that to let students feel online education can provide higher teaching quality level to compare traditional classroom learning method.

Other potential impacts of education market method relates more to education going online and a shift away from the more traditional forms of campus-based teaching in highest education . Would improving access to online education have the effect of increasing online students number. Due to who accept to choose online learning from traditional classroom learning habits . Thus, this is one learning habit change challenge for the traditional classroom learning students to adopt the online learning habit change.

In conclusion, for online education providers who need to consider how to change traditional classroom learning and teaching habit to adapt new online learning and teaching habit, as well as how to provide online teaching quality is higher level to compare to traditional teaching quality if who want their online education service businesses are successful.

- Future innovative and sustainable food source market

Future human considers health , so whose demand will high for quality of foods, farming of fish, typically freshwater with the cultivation of plants. It is simple future food source needs high health quality to provide to human to eat. If the food manufacture can have method to innovate any food quality to be more health to reduce poor health risk to influence human to eat. Thus, the food manufacturing process will be one important factor to attract consumers to choose to buy the food manufacturer's food supply. So, health food source market must attract many consumers to choose to buy to eat.

Aquaponic system will be one health food manufacturing method. Aquaponic systems combine the farming of fish, typically freshwater, with the cultivation of plants. This takes place within a closed -loop aquaculture system, whereby fish are fed nutrients and their excrements one need as fertilizer directly into the water in which they are being loop. The water then feeds plants which use it for growth and filter the water , so it is suitable for re-use with the fish in the system. Such a system can be said to be closed-loop and hence a significant emphasis is placed upon the environmental and economic sustainability characteristics of acquaponic systems are only small-scale and therefore incur high costs of production relative to current methods of large-scale-farming.

However, in the future, due to human ought consider health, so who we need any food have good health quality to avoid any illness, e.g. cancer, even death causing risk from bad health foods source. In conclusion, food manufacturers need to consider any new food manufacturing methods to achieve how to manufacture foods to keep fresh and health level if they hope their food products can be attractive to consumers to choose to buy to

eat.
Hence, future any new technological invention to manufacture health food will be one important factor to influence global food industry development. Also, it implies any food manufacturers need to consider how to manufacture any health foods in whose food manufacturing process. In conclusion, future foods and agriculture development will be trend to agriculture health productivity, avoidance from pests and diseases influence to food manufacturing process, avoidance food supply inequality and insecurity, more nutrition and health, changing food source manufacturing process system, reducing food losses and waste during food manufacturing process, making food systems more efficient, building resilience to protracted arises, disasters and conflicts, preventing transboundary and emerging agriculture and food system threats.

Future business strategy trends

- Government (public) and private partnership property development strategy

In future some business, public and private partnership method is more suitable to compare the private entrepreneur sole operation. For example, property development, construction industry example, building and rebuilding cities and new communities is a complex challenge, it requires public and private interests and resources. However, the traditional process of urban and suburban can be developed between the local government and private property developer, which will win distinctly different benefits if they decide to cooperate together.
The need to rebuild and revitalize older portions of urban areas , the public need to monetize underused assets have dramatically changes. In fact, private sole property

developer's disadvantages is that it has no longer can private capital be relied on to pay the high price of assembling and preparing appropriate sites for redevelopment. Also, it has no longer can local governments bear the full burden of paying the costs of public infrastructure and facilities. If public housing department and private property developer can cooperate to achieve shared goals and objectives, this process can require applying far more effort and skill to weighing, and then balancing, public and private interests and minimizing conflicts.

For another public and private partnership example, such as health care providers and education institutions, non profit associations, such as community based organizations and business improvement district organization , these organizations are very suitable to choose to cooperate with government (public organization) to do their businesses together in the future global business environment trend.

However, the property development industry will have more benefits and needs to choose public and private partnership to compare these above industry. The main reason is that this industry will much capital to invest to any building business and it is long term tangible fixed property development business. Thus, the public and private property development partnership can implement a range of pursuits from projects to long term-plans for land use and economic growth. Partnerships have completed real estate projects, such as mixed-use developments, urban renewal through land and property assembly, public facilities, such as convention centers and airports and public services , such as affordable and military housing.

However, each public and private property development partnership is the best to share common stages with each

development process as below:
In the first stage, conceptualization and initiation, stakeholders' opinions of the vision and surveyed and partners are selected through a competitive process.
In the second phase, entities document the partnership and begin to define project elements, roles and responsibilities, risks and rewards and the decision and implementation process.
In the third phase, the partnership attempts to obtain support from all stakeholders, including civil groups, local government (through entitlement), and project team members.
Finally, in the fourth phase, the partnership begins construction, leasing and occupancy and property and asset management.
However, the process is repetitions and can continue beyond the final phase when partners manage properties or initiate new projects.
For US one successful public and private property development partnership example, the contributing major benefits to the citizens of Washington, D.C. The James Foyster School Henry Adams House, a public elementary school and 211 unit residential apartment complex was constructed as a result of a partnership among the District Of Columbia Public Schools.

- Tourism partnership

Another a major public and private partnership is tourism industry. Will public and private partnership to tourism be better than sole travel agent business operation? I believe it is better to any travel agent to choose public and private partnership strategy, the reasons include as below:

- Public tourism partnership goal, to provide the countries' different tourism destinations and tourism features to consider any country tourism information to assist travelers in understanding the travel problem, alternatives, opportunities and/or solutions to adapt every traveler individual travelling need.
- To obtain public travelling feedback on analysis, alternatives and/or travelling decisions, to work directly with the public throughout the travel public promotion process to ensure that public concerns and aspirations are consistent understood and considered, to partner with the public in each aspect of any tourism tickets comparison, tourism destinations, tourism entertainment, and transportation of the decision information , including the different tourism destinations development of alternatives and the identification of the preferred solution to place final decision -makings in the hands of the public.
- Every travel agent and public organization will keep every traveler's informed, listen and knowledge concerns and provide feedback on how public input influenced every tourism decision to every individual traveler considerately.
- Tourism techniques will consider fact sheet, web sites, open houses, public comment, focus tourism groups, surveys, public meetings, workshops, deliberate polling.

In conclusion, public and private partnership will be future trend to develop because capital can be shared, reduces sole business operation risks, promotes business information efficiently and easily when public organization can participate to assist these private business organization to cooperate to develop their businesses together.

- Higher education marketing, enrollment, branding and recruitment strategy

Future the most important tools for social and online education marketing will be an effective university website promotion tool to build ultimate brand for any university organization. Websites often feature elements and highlight content, including navigation, bars, engaging visuals, such as slideshows, and prominent " call to action" buttons that encourage students to apply. For example, radio ads., asking current students or for applicant referrals and online college fairs were deemed least effective, when the most effective methods of outreach open houses and campuses visit days for high school students.

Online education courses will be popular, due to adaptive learning technology has also enjoyed. So, successful branding can help increasing enrollment, expanding fundraising capabilities and other outcomes. Today, effective strategy planning and brand management require more than traditional advertising. Education institutions present and manage brand message, experience and environment achieve a competitive advantage in recruiting, building royalty among their students, parents , staff , faculty and donors.

In conclusion, how to do effective website advertising to promote university courses, website enrollment method? I shall recommend these methods as below:

Firstly, design responsive website, education institutions are placing more emphasis on responsive web design to create intuitive and easy to navigate websites that can be viewed on multiple, devices and platform.

Secondly, university administrators want their education institutions to receive a spot in search engine result particularly Google website. Especially for education institutions that offer niche programs , it is increasingly important to ensure that search results, including the programs at the top.

Thirdly, how to use of web analytics, colleges and universities are relying on data-driven analytic to determine who, whom and where they are reaching their audiences. The use of analytics software is increasing as the higher education web ecosystem is becoming complex, e.g. domains, subdomains etc.

Fourthly, getting a better handle of this data is a new area of concentration for colleges and universities strategic social media, when recent polls indicate nearly every education institutes of higher education use some form of social media, e.g. face book or twitter account, these trends are explored.

Fifthly, the rise of mobile development and connected decides to colleges and universities for a greater amount of course content of mobile versions of websites to promote to every student to know from whose every mobile, CRM systems are heavily on content management and customer relation systems for admission for prospective students service in the future mobile promotion technology.

CHAPTER IV

Artificial intelligence positive or negative influence

1.1 Will artificial intelligent technology cause war?

Nowadays, artificial intelligent technology is developed to be beneficial to bring positive impact to influence human's life. It is one kind of automatic machine learning to help human to seek the best solution to any problems by trial error to sure how the result is reached, it has computing power to learn any new skills, it is created artificial intelligence to own human mind to exceed ourselves human intelligence as well as the creation of new types of jobs. Such as non-manual driving automatic vehicles, insurance agents, investment analysts, and medical diagnostician etc. different trial error and mind analytical needs of characteristics jobs. Thus, it seems artificial intelligence will be possible to invent to the stage of own human mind. If their mind is bad, it is possible that which can dominate our life in the future.

Moreover, scientists predicted up to 50 % of today's jobs will be lost to these trends. Consequently, it will raise global unemployment number in the future. Thus, it seems that it will influence much human natural jobs are replaced by artificial intelligent job natural jobs in the future one day in possible. So, it means that (AI) can be perhaps to dominate human's mind and behavior in our daily life.

Thus, it also brings one challenge: Will (AI) threaten human's society to cause life danger if (AI) technology is applied to attack human to cause war by some countries‘ ambitious leaders. It is clear that this is a strange human’s aim or intention to apply to (AI) technology, because it depends on what the (AI) manufacturers and/or inventors and/or the country’s leader to consider being intelligent in the behavior of how they intend to apply (AI) technology in the future.

Thus, if any (AI) technology new discoveries are applied to assist employers to raise workers’ productivities or performances and to help human to find error and trial and solve challenges. Then, it will bring benefits to human. Otherwise, of (AI) technology are applied to dominate human behaviors or minds or damage world peace. Then, it is possible that to bring technological war among ourselves. However, to achieve whom will be the leaders of our world’s technological dominance. Thus, it is a horror technological war to influence our daily life in the future one day.

1.2 Reasons (AI) technology can bring negative influences

Why human will apply (AI) technology to contribute negative influences? What factors causes their negative behaviors? I shall give my opinions to support my prediction, judgement and uncertainty dominant of (AI) technology damage and to dominate to future human’s societies that it is possible to occur.

Firstly, on artificial intelligent improvements in prediction (AI) technology and immoral decision making reason there is a risky action to scientists , whose payoff depends on a safe action with the same payoff in every application of situation. Judgement is costly, for each potential application, it requires thought on what the payoff might be. Prediction and judgement are complements as long as judgement is not too difficult. If some (AI) inventions are judged to make immoral decision to be applied in damage human safety aspect by scientists. Can we control it? It is the world's greatest opportunity and its greatest threat to encourage or attract the (AI) scientists to choose to do any technological damage behaviors to cause war occurrence indirectly.

It seems that mathematicians, philosophers, computer scientists and engineers who have responsibilities to spend their days thinking about how to avert catastrophes: meteor strikes, nuclear winter, environmental destruction, threats, when who attempt to apply (AI) technology to cause these damages to our earth.

There are a lot of things, that can go and have gone wrong throughout history, such as earthquakes and wars. But, these is one kind of thing that has not ever gone wrong. It is permanently destroyed the entire future by (AI) technology misuse. Thus, such as a man-made one: that rapidly advancing research into artificial intelligence might led to a runaway " superintelligence" which could threaten our survival. Whether countries‘ leaders will apply (AI) technology made machine men (machine soldiers) to become such as human soldiers to attack other countries in the future technological military war gaming. It means that if one day (AI) technological invention can implement to manufacture machine made soldiers successfully, then it can give chance to the ambitious countries' leaders to apply them to replace human soldiers to attack other countries. Then, the technological wars will be caused in possible in the future one day. Thus, scientists need to consider the future of defense to avoid artificial intelligent technology is controlled or dominated by the ambitious countries‘

leaders.

Against (AI) armed forces themselves. There are the active agents in terms of wetware (humans), hardware and software behind our efforts, generate defense and security efforts. Some scientists indicate the general artificial intelligent level had been improved to the artificial super intelligent level. Then, the technological war threats will be raised. Then, any countries' defense providers need to concern to predict when artificial super intelligence will achieve to this level to be caused technological war by (AI) artificial super machine soldier inventions.

1.3 The future of weaponized artificial intelligence

Why does it possible artificial intelligenc is weaponized? There are three key threat areas regarding of weaponized (AI):

- (AI) surveillance
- The (AI) weapons factory
- Careless destabilization of national security

Human future threat casting uses inputs from social science, technical research, cultural history, economics trends. We need to concern (AI) threat challenges, such as: Can we develop (AI) and rethink th very nture of (AI) without losing control over it? How to approach threatcasting and future modeling from an economic perspective? What will be the growth , impact and future of applying (AI) to real world industries?

(AI) threatcasting is a theoretical exercise undertaken by knowledge practitioners, such as (AI) scientists with special domain knowledge of how to specifically discrupt, mitigate, and recover from theoretical threat futures. Thus, (AI) threatcsting can be applied on technological military war problem. If human can predict when (AI) technology impacts to our lifes. We can predict when the future we want from (AI) technological achievement and the future we want to avoid from (AI) technologicl war.

1.4 What is (AI) weapons factory?

When one country has enough funding and the support to develop its most ambitious object to date, a super (AI) that could manage the world's energy and climat change. It is horror that it was actually building the world's largest (AI) weapons factory with the capability to invade every country in the region . Thus, it is only developed technological country, it has effort to invent (AI) super weapons to dominat or control overall world. If it aims to apply (AI) technology to give welfares to help human to solve challenges, e.g. climat change, seeking shortage enery. It is a good matter. Otherwise, if it aims to apply (AI) technology to dominate human's life safety. It is not a good matter. Thus, scientists need to avoid to do immoral behaviors to build (AI) weapons factory.

1.4.1 What is (AI) weapons?

In our future, now scientists will apply traditional (AI) technologies to manufacture (AI) weapons. Experience with traditional weapons allow organizations to understand the immediate mortal threat. (AI) weaponry shifts armament into more difficult to track, more integrated and systematically impactful.

This, is not only applies to individuals , but also to systems on an unprecedented scale. Imagine the destruction of an entire city energy system or the turning of common household connected devices (internet of things) into actors. As (AI) continues to be integrated into everyday tasks as well as into core everyday tasks as well as into core functionalities of cities and governments, the potential for turncoat or altered (AI) rises, increasing the potential for integrated (AI) threat actors operating behind the scences. Thus, the weaponization of (AI) presents a new challenge as we imagine the changing nature of factories where software , instead of hardwares are created the crime have challenged how organizations defend and protect themselves. The coming weapons factories of (AI) will present a whole ethical, legislative and security issues concerning.

How to define and locate (AI) weapons factories? I feel the (AI) factories are no longer solely buildings , but a mix of vitual and substantially different facilities. Particularly as it shifts from a physical assemly and development model to a distributed and flexible networks. Needing minimal raw materials to develop weapons, the physical location of these factories could be anywhere and their identification from the outside, nearly impossible. To conclude for both benefits and threats, the integration of (AI) will impact the vast majority of human, changing how we interact, work and live. It means (AI) weapons are one kind of high technologicl internet skillful threat to our daily life. Thus, we need to consider hoe internet development to avoid netgative influence to our lifes.

If (AI) artificial intelligent digital systems are invented, when we use this new (AI) technology, we need o aware these questions to avoid (AI) digital systems threats. Such as how your personal data is being used and the implications , both positive and negative of sharing data. Demand that brand and organizations practice are inform you of how they are using your data. Awareness with populations and communities without access to training or education about (AI) safely. Explore the creation of an international organization that can oversee the development of (AI) to ensure that it is not weaponized.

1.5 Why (AI) technology will be dangerous?

We need to know technology is not science. Scientists are perceived as middle-aged, emotionally impaired and dangerous human. Science tells us how to the world is . That we are not at the centre of the universe is neither good nor bad, nor is the possibility that genes can influence our intelligence or our behavior. Dangers and ethical issues only when science is applied as technology . However, ethical issues can arise an actually doing the scientific research. Thus, it ws imaginative trial and error, such as (AI) technology that carries with it ethical issues from motor cars to polluting the environment and (AI) technology weapons of war in possible.

Some which the nucler was obtained. It seems scientists also do any hurt natural life behaviors to achieve experimental successful aim. Such as (AI) technological invention can be either beneficial or harmful to human's safety if who decided to profitability aim.

In conclusion, artificial intelligence is weaponized, it is possible to cause, it depends on the (AI) scientists to choose how to implement their (AI) scientifical rescarch mission . Considerately, when general (AI) technology will be developed to reach super (AI) technology stage, human's life safety will be also threatened by super (AI) technology. Thus, (AI) scientists need to concern their behaviors whether are moral.

Printed by Libri Plureos GmbH in Hamburg,
Germany